Gary the Glo...

Name _____

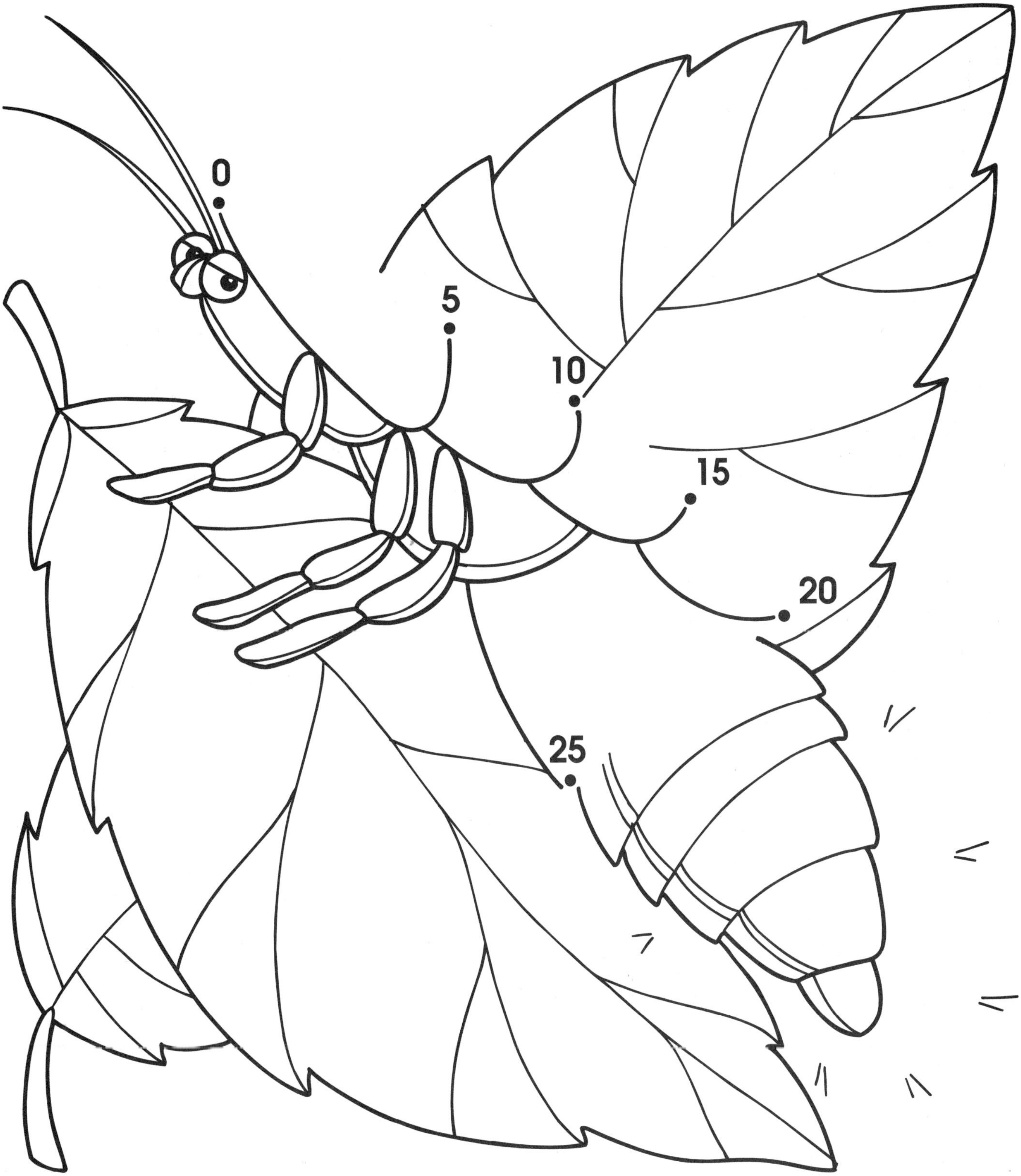

Angie the Ant

Name _____ Skip counting 0–30

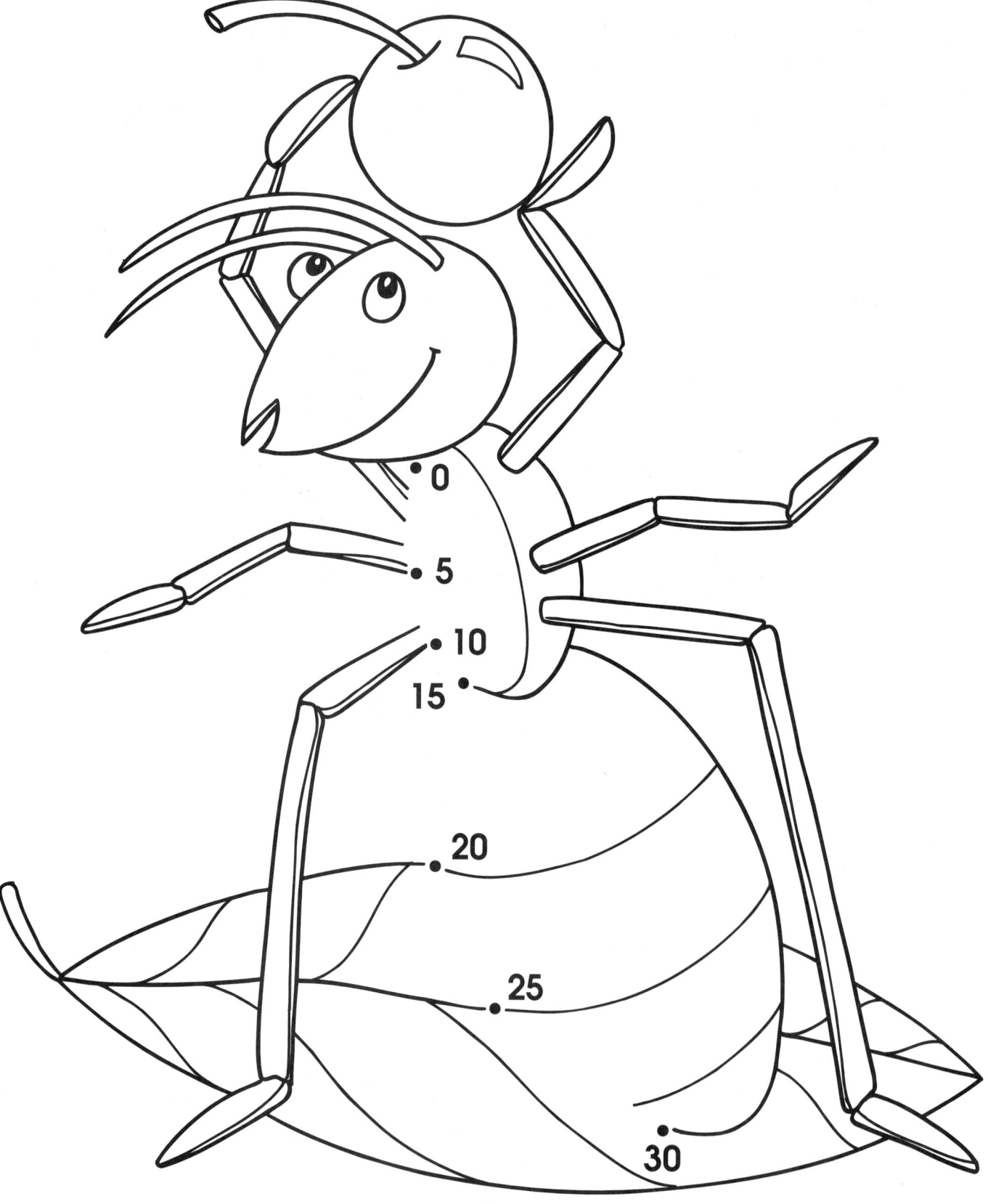

Thomas the Thorn Bug

Name _____

Skip counting 0–35

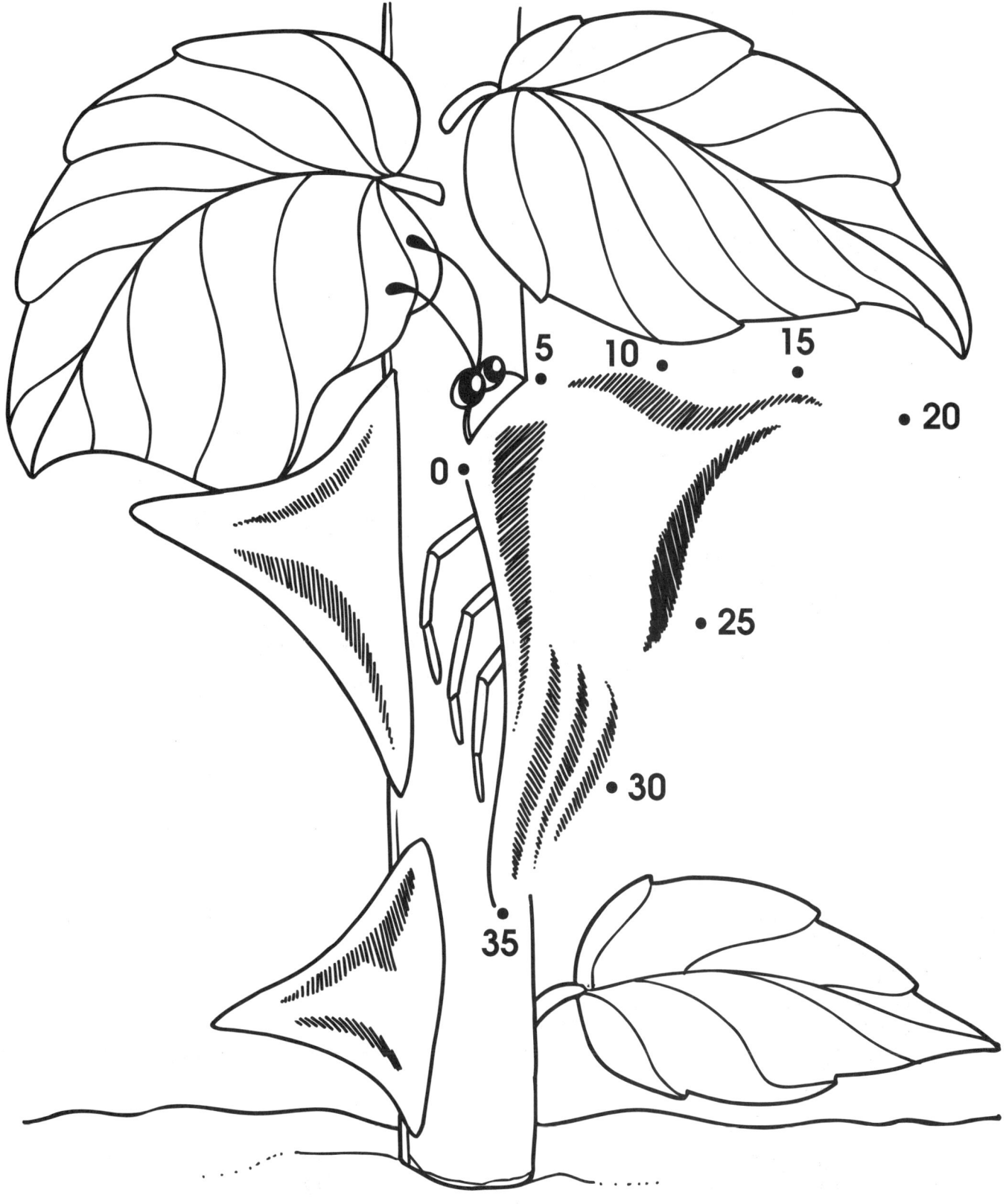

© Frank Schaffer Publications, Inc.

FS130308 Bugs

Mary the Millipede

Name _____ Skip counting 0–40

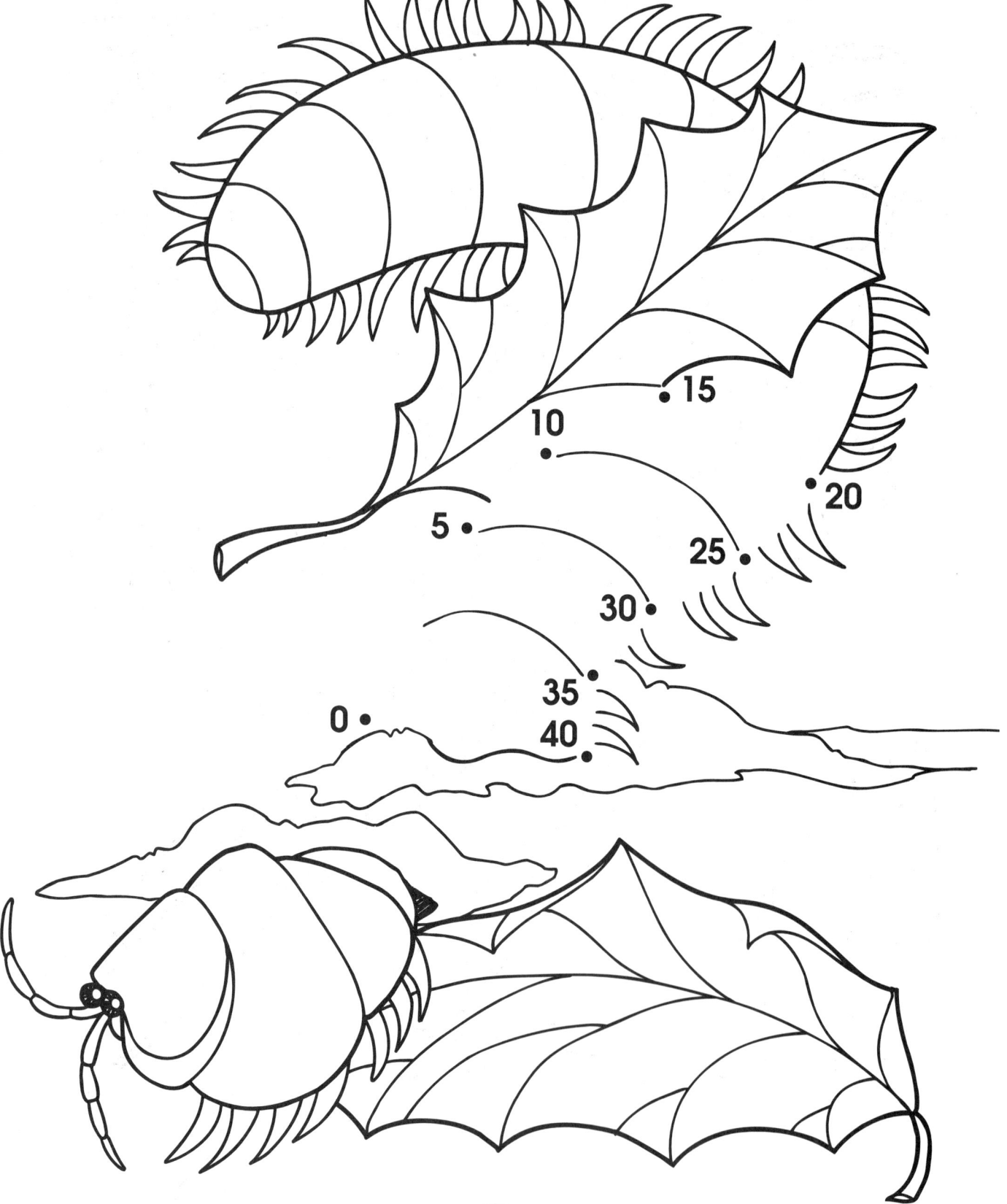

4 reproducible

Willy the Walkingstick

Name _____ Skip counting 0–50

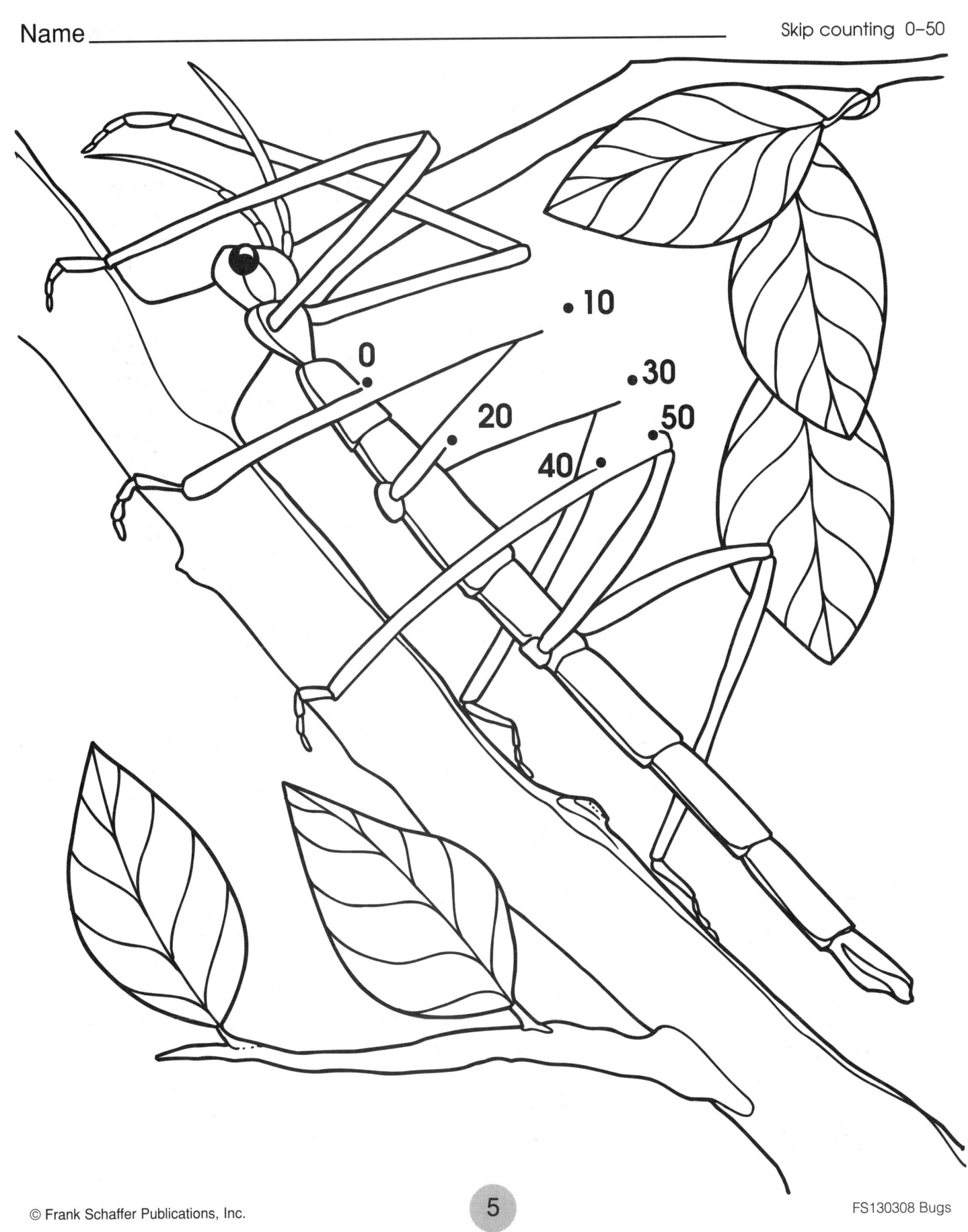

Amy the Atlas Moth

Name _____

Skip counting 0–60

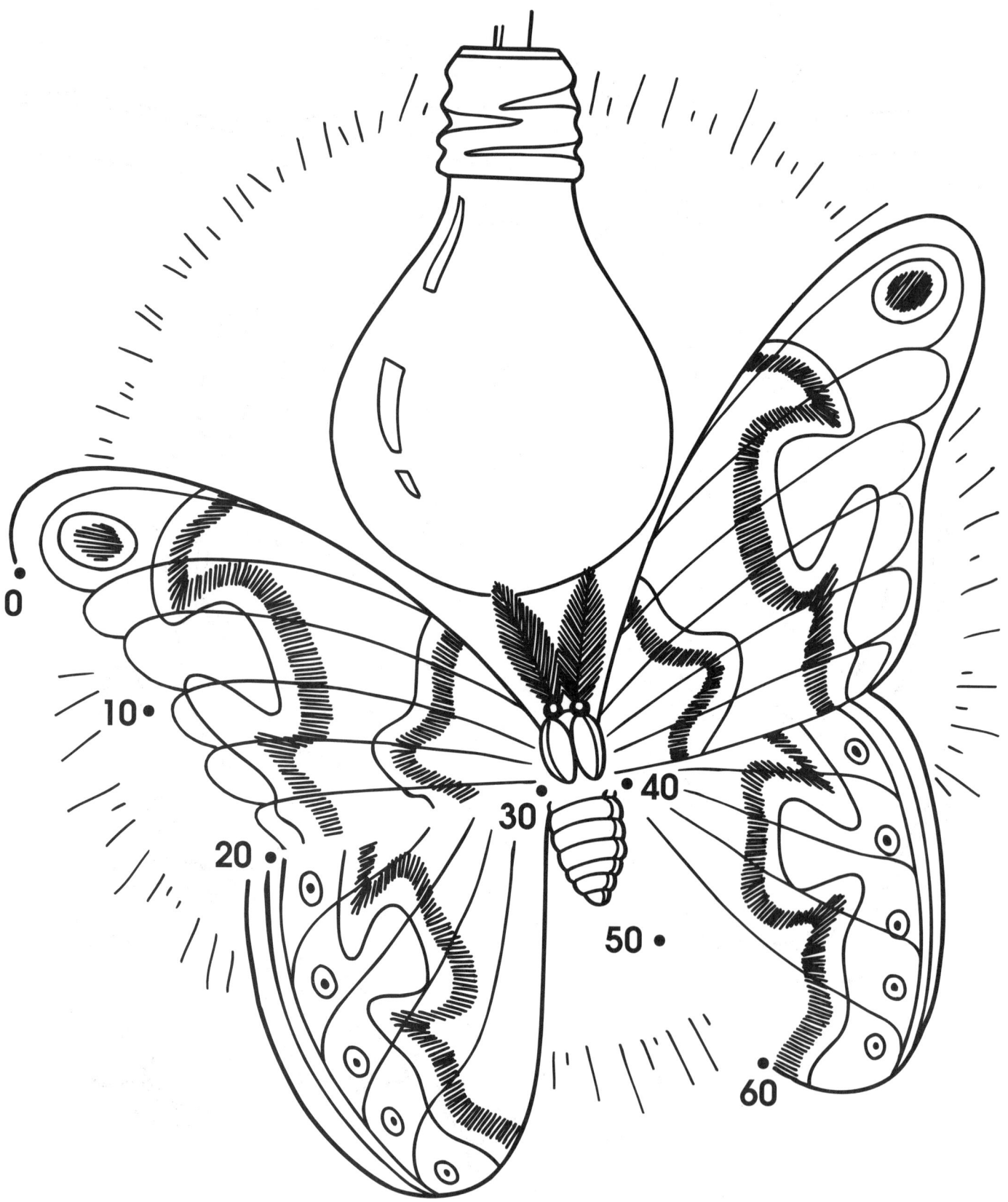

Walter the Water Bug

Name _____ Skip counting 0–70

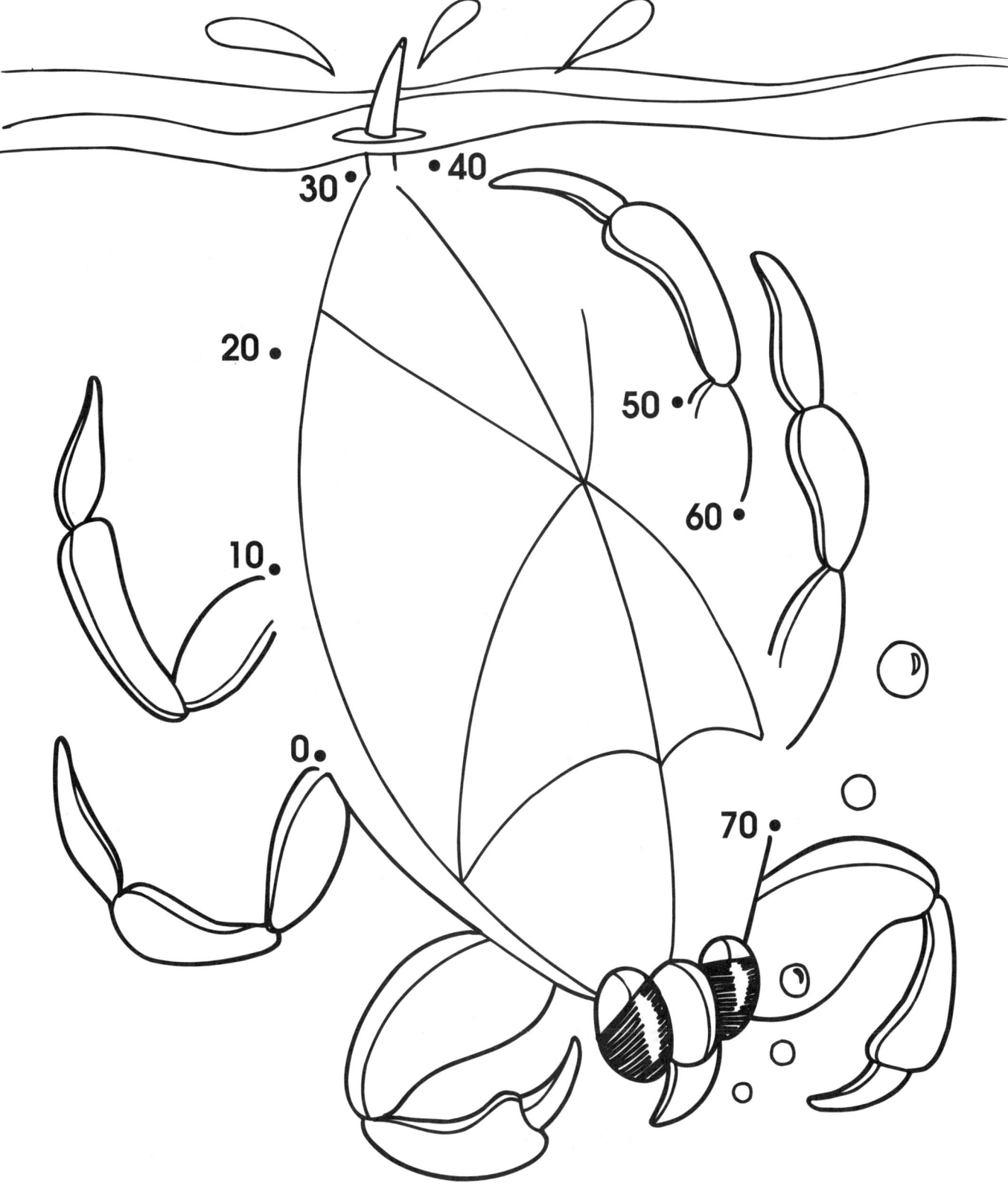

Marty the Mantis

Name _____

Skip counting 10–80

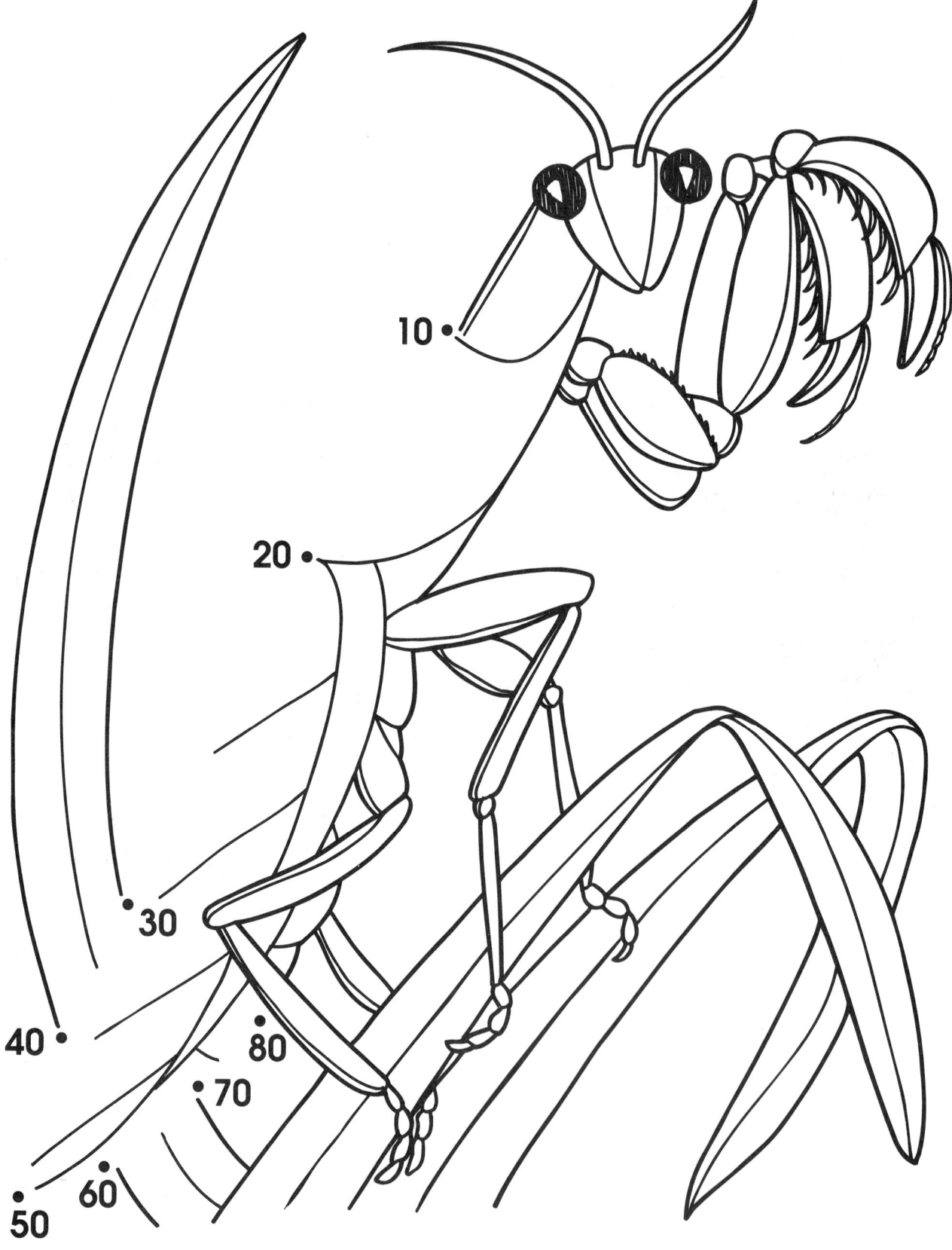

© Frank Schaffer Publications, Inc.

FS130308 Bugs

David the Dragonfly

Name _____

Skip counting 0–45

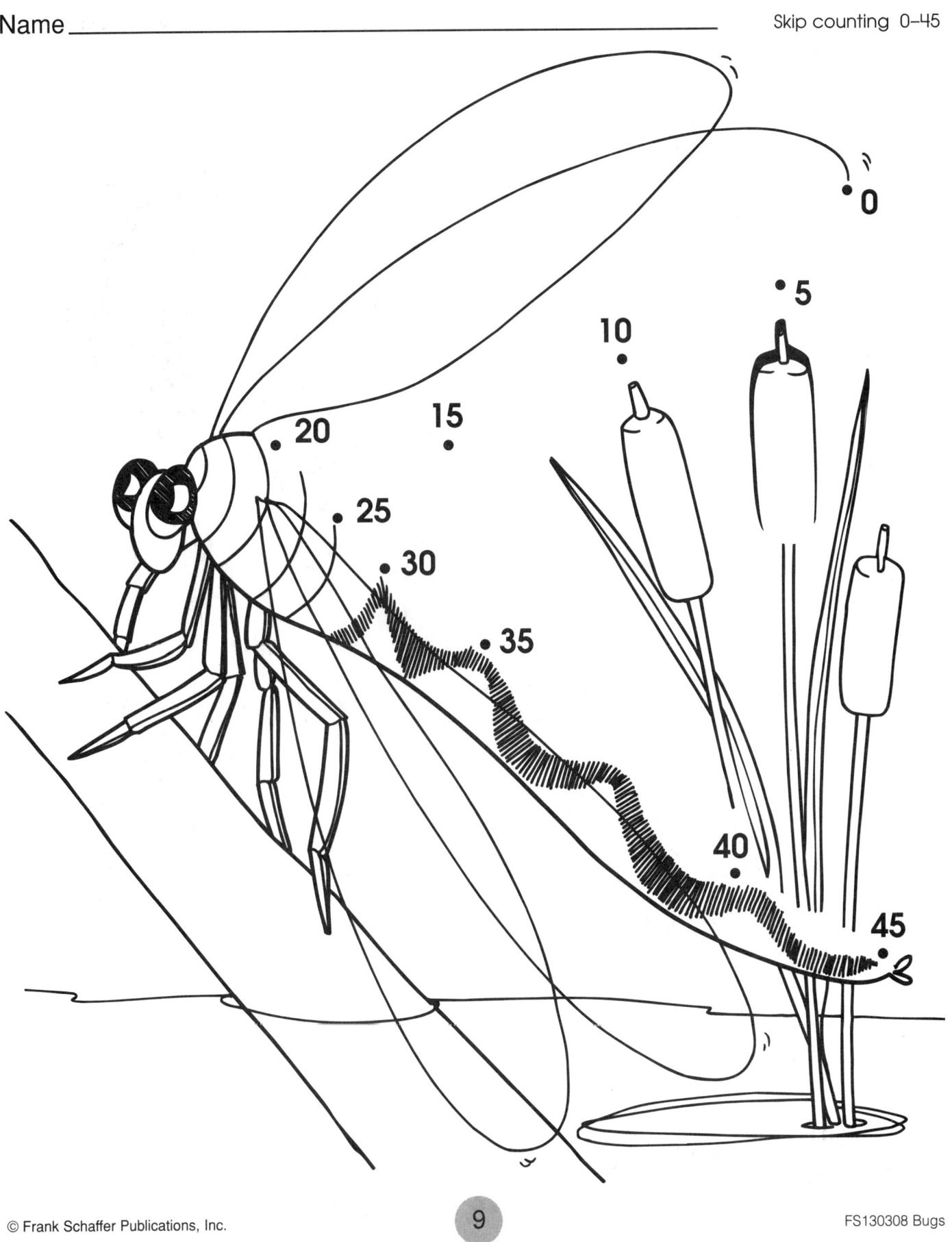

Gina the Goliath Beetle

Name _____

Skip counting 5–50

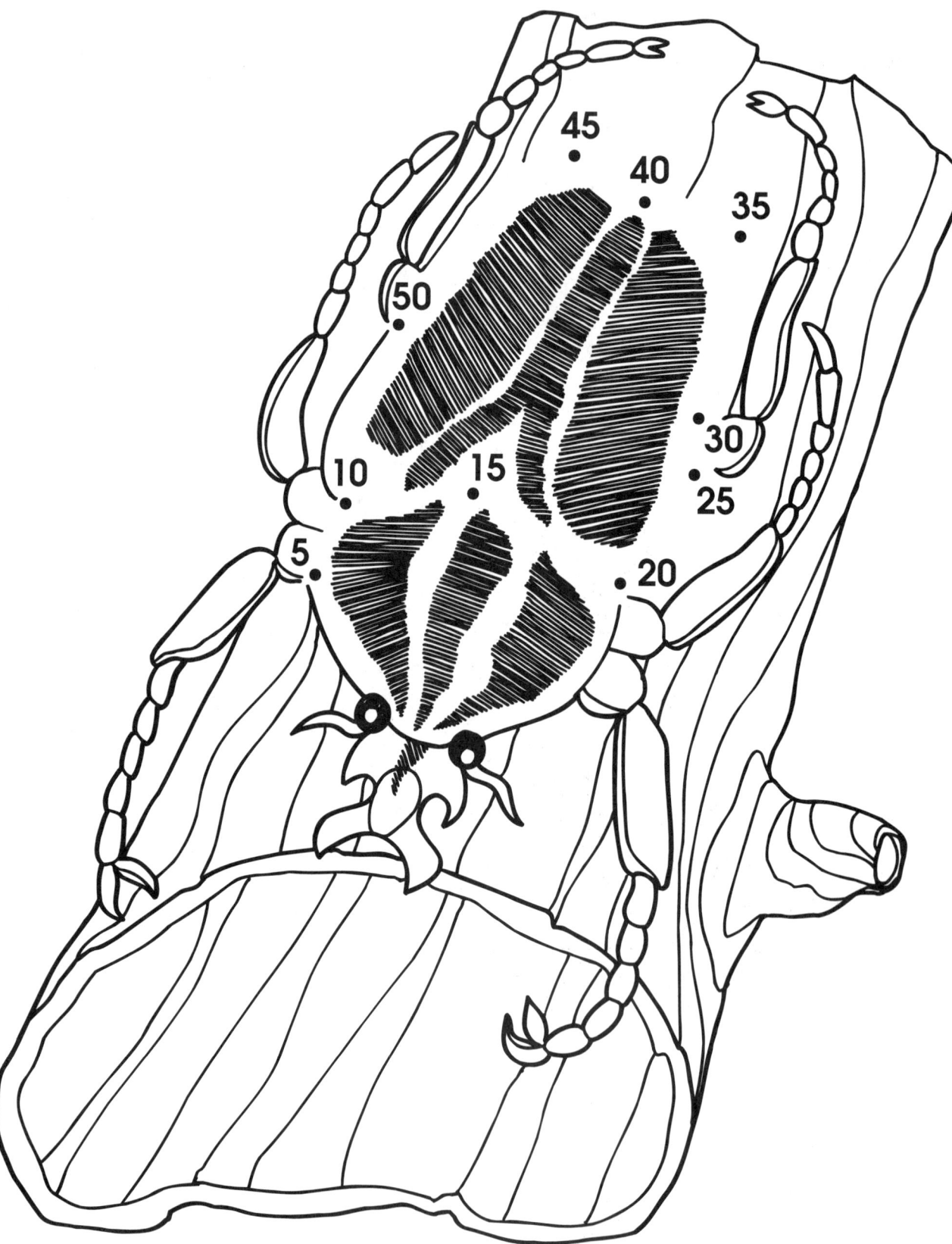

Buddy the Bee

Name _____

Skip counting 0–60

30 •
35 •
• 60
25 •
• 40
55 •
20 •
15
10 •
45
• 50
5 •
0 •

Vera the Velvet Ant

Name _____ Skip counting 5–65

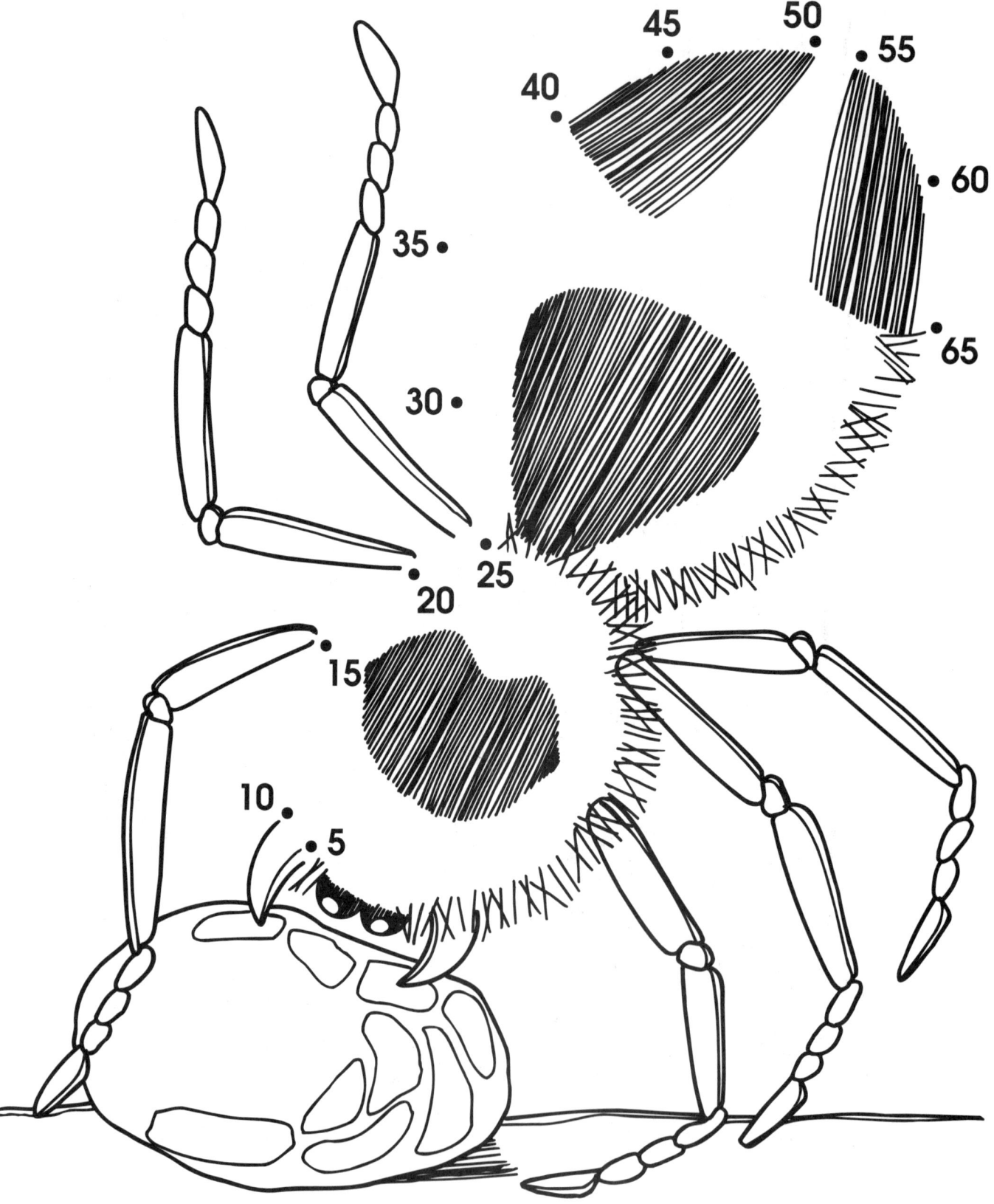

Oscar the Owlet Moth

Name _____

Skip counting 0–90

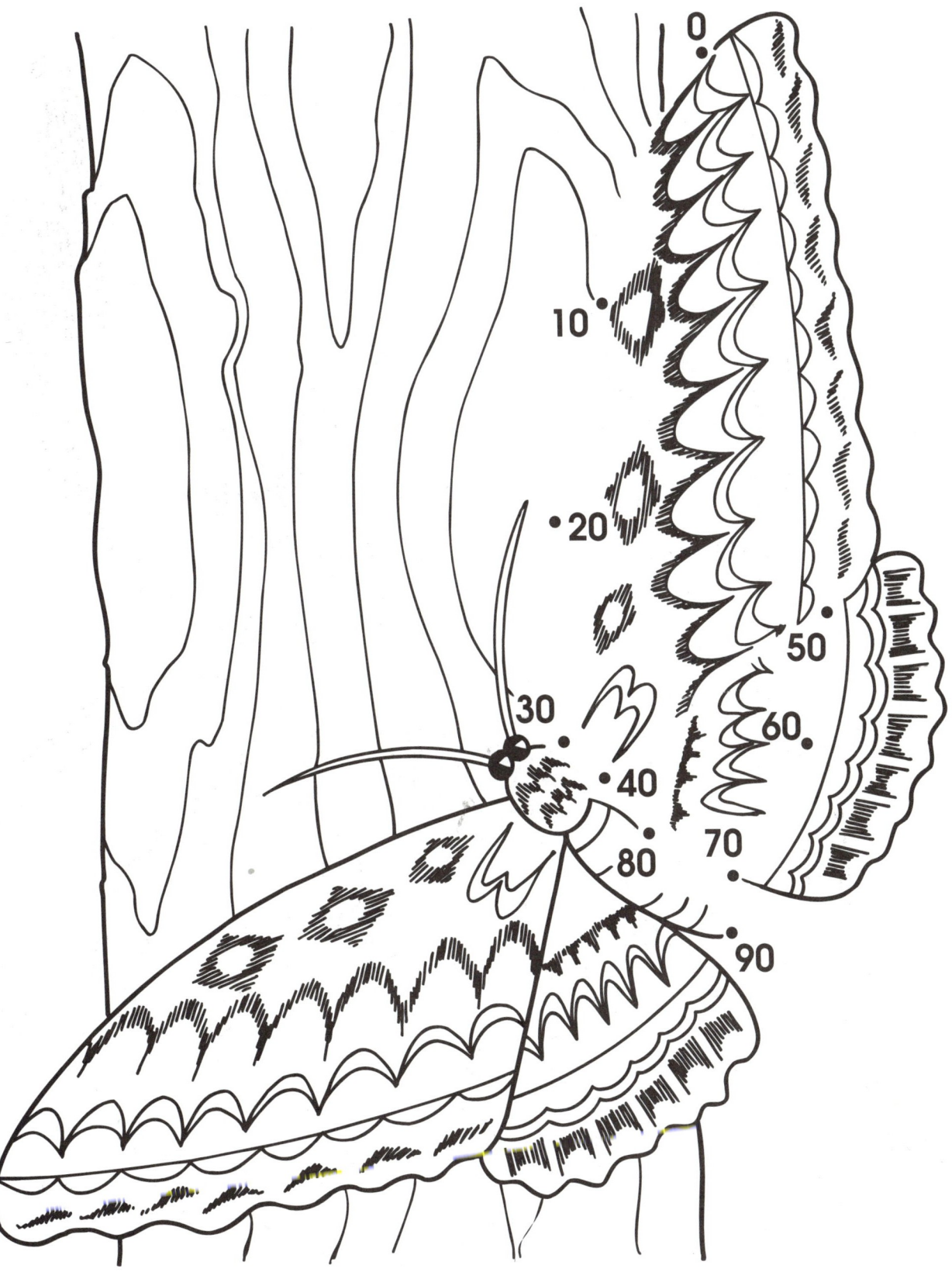

Lily the Ladybug

Name_____ Skip counting 10–100

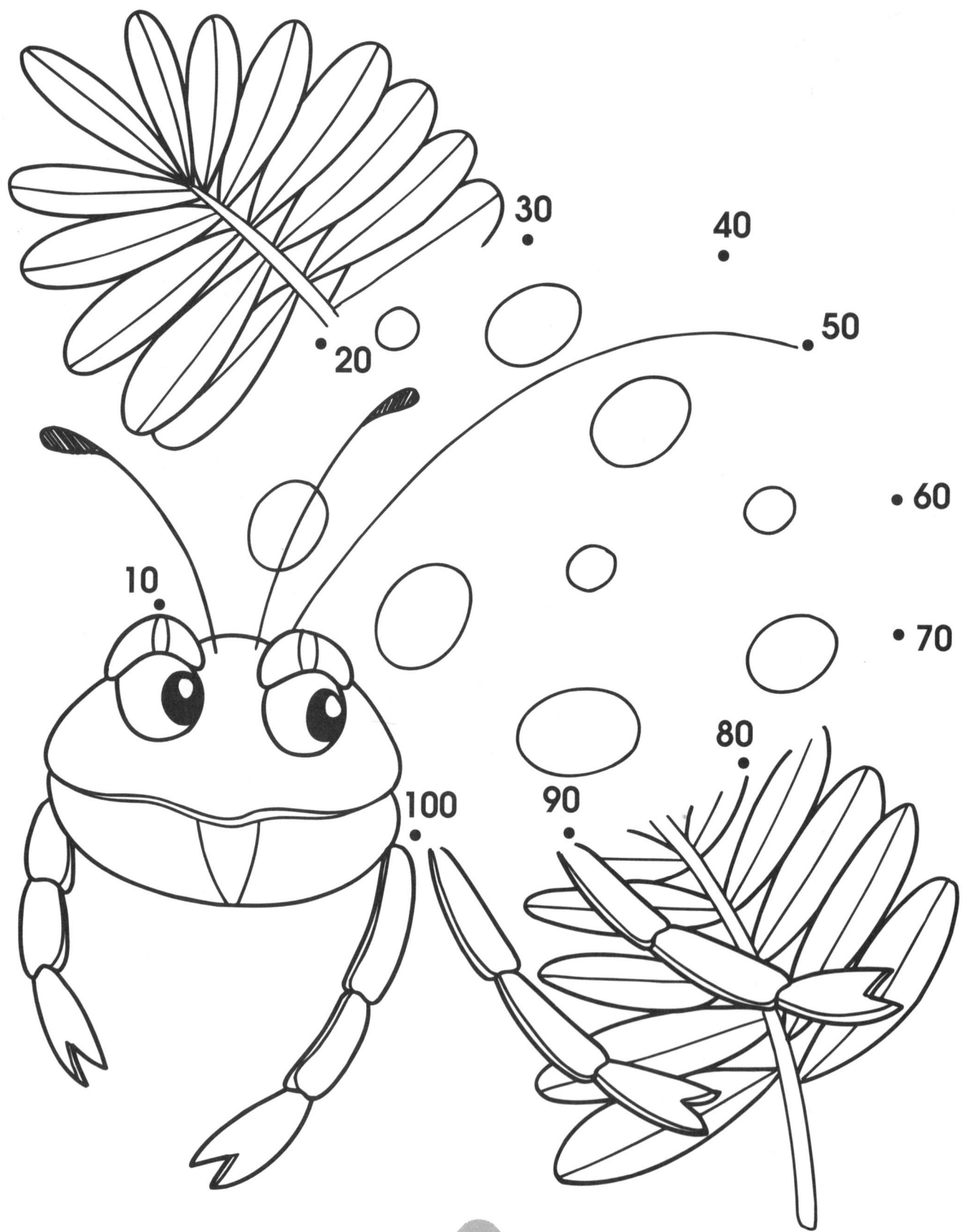

Manny the Mosquito

Name _____ Skip counting 0-110

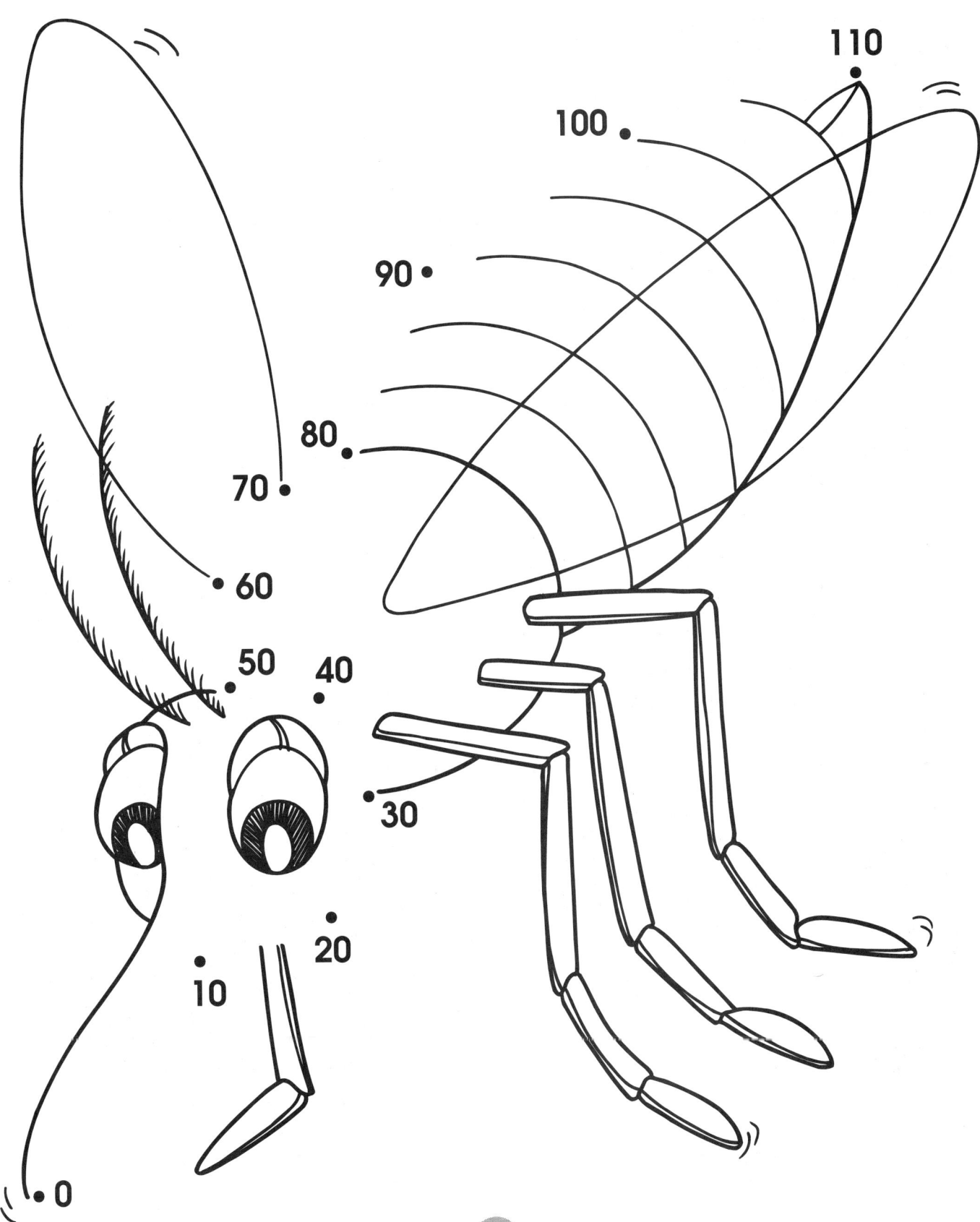

Katie the Katydid

Name _____

Skip counting 10-120

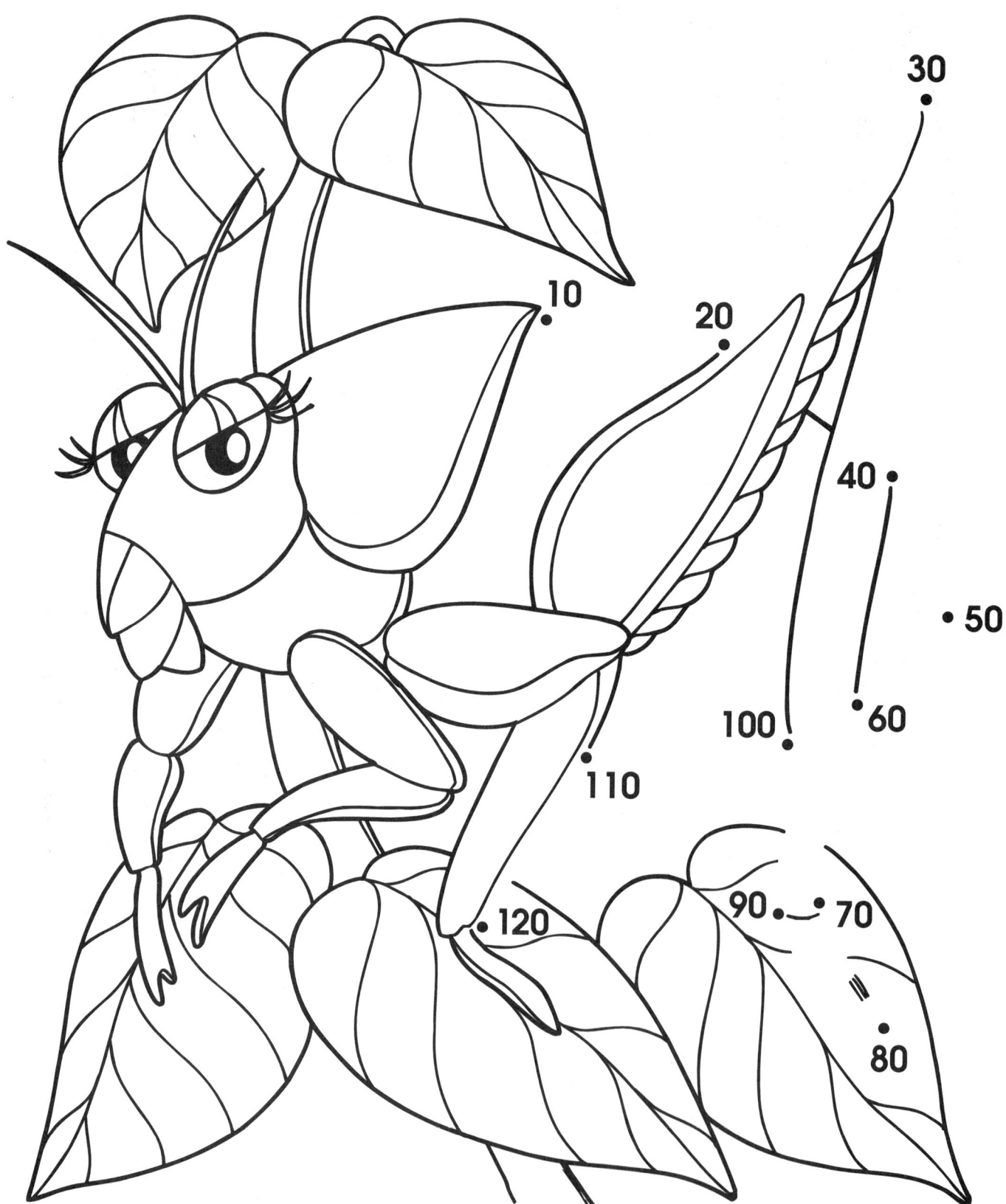

Taylor the Treehopper

Name _____

Skip counting 0–65

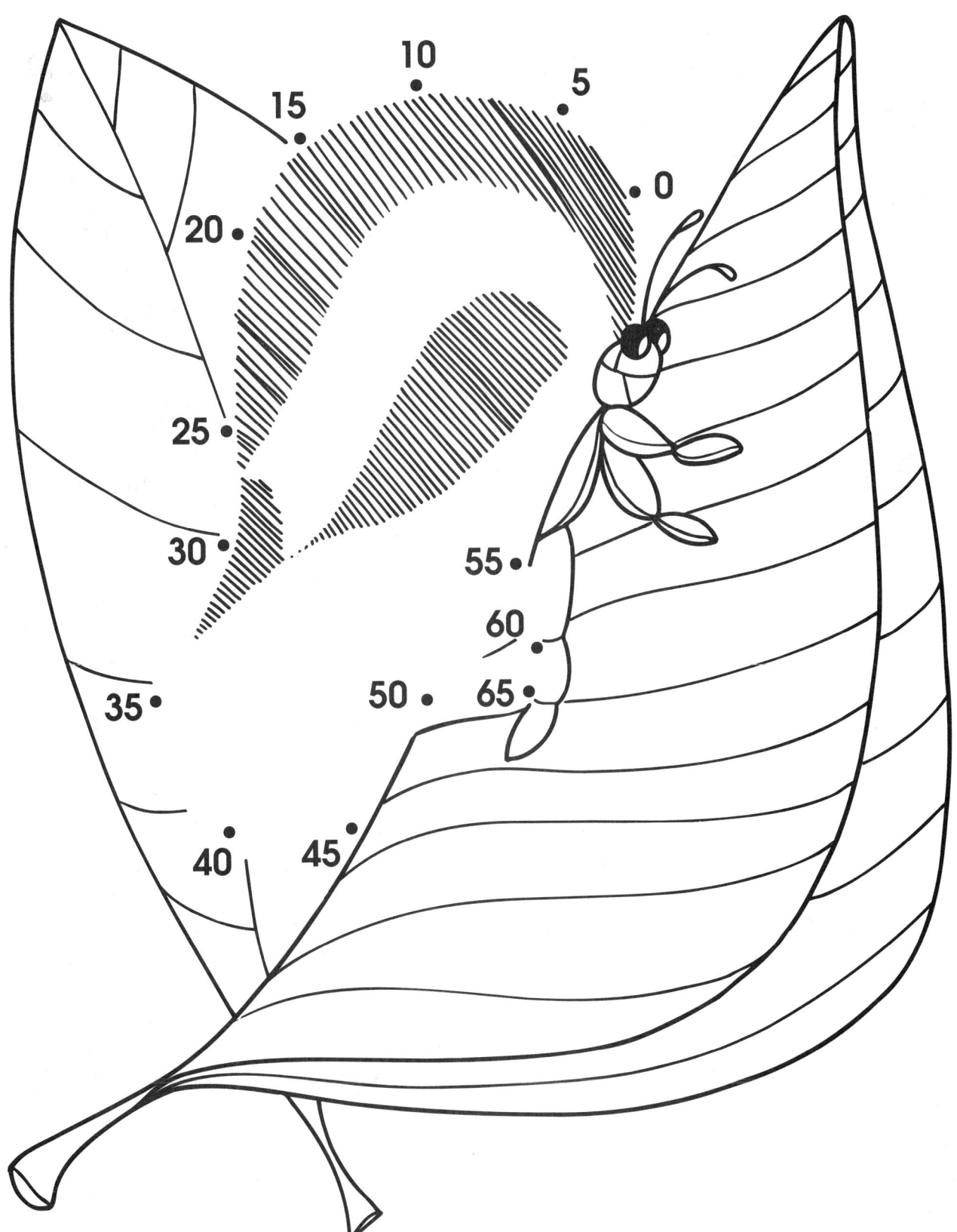

Violet the Violin Beetle

Name _____

Skip counting 10–75

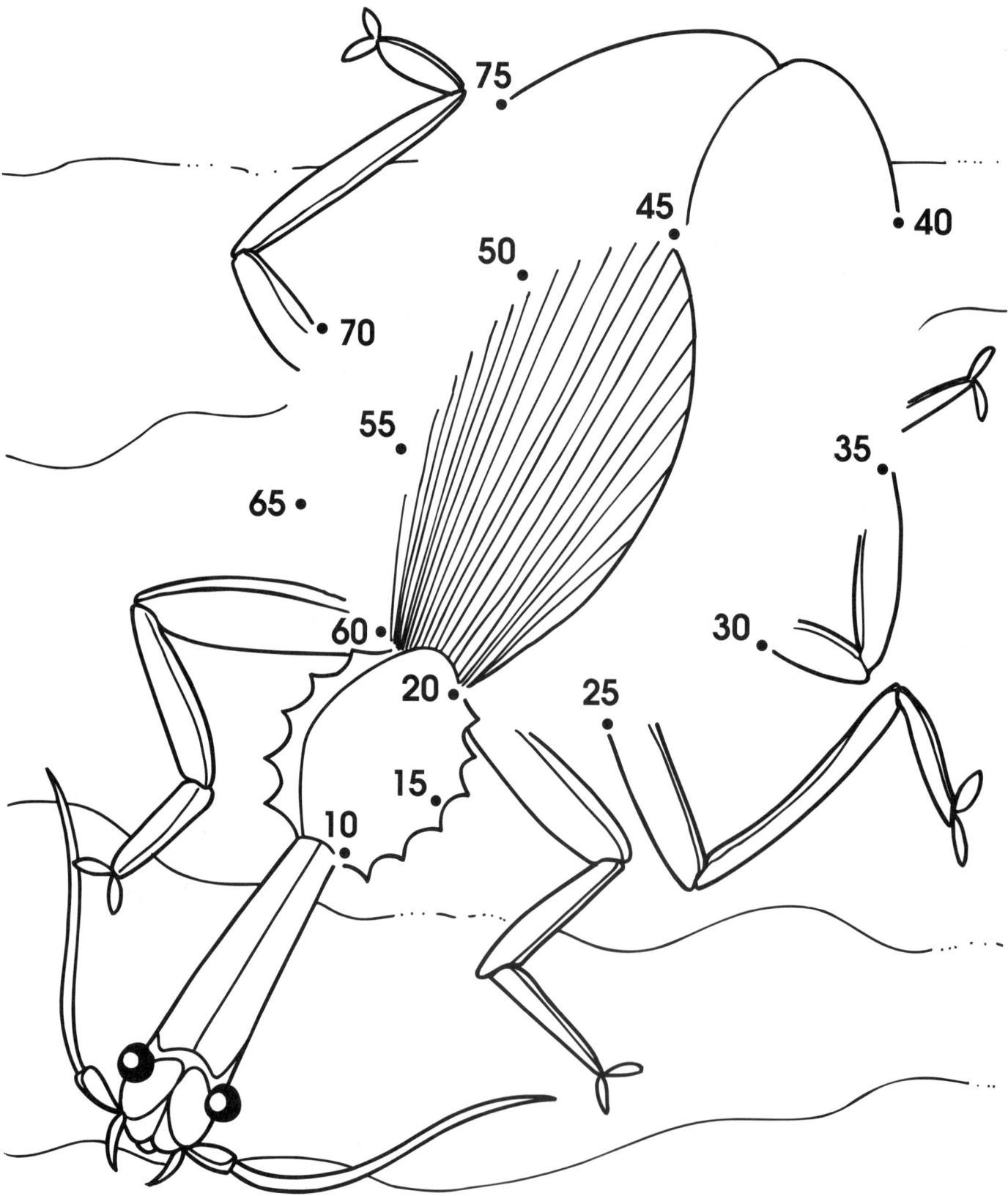

© Frank Schaffer Publications, Inc.

FS130308 Bugs

Yoli the Yellowjacket

Name _____   Skip counting 15–90

Irene the Io Moth

Name _____ Skip counting 25–100

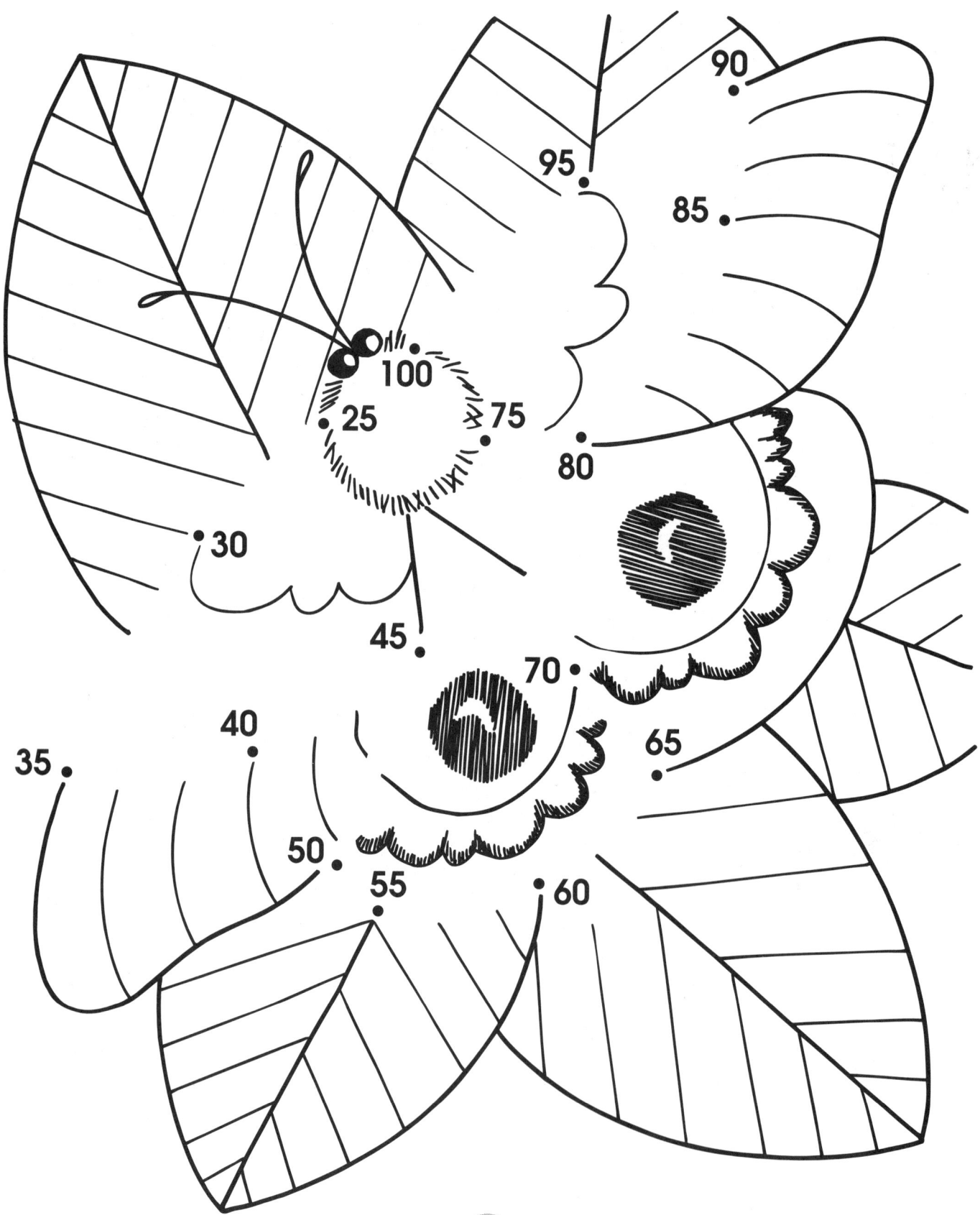

Doug the Darkling Beetle

Name _____ Skip counting 0–130

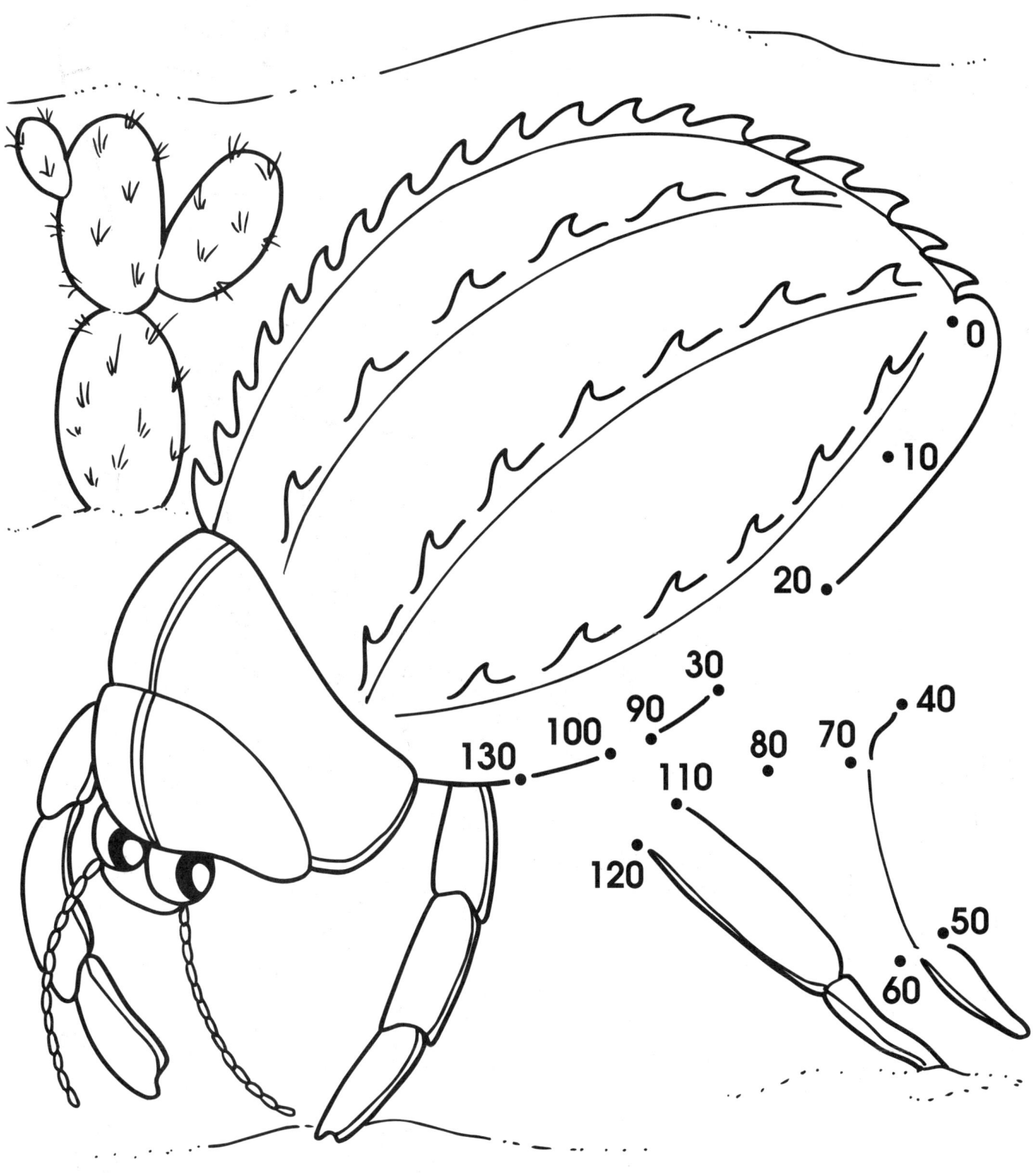

Beth the Butterfly

Name _____

Skip counting 40–170

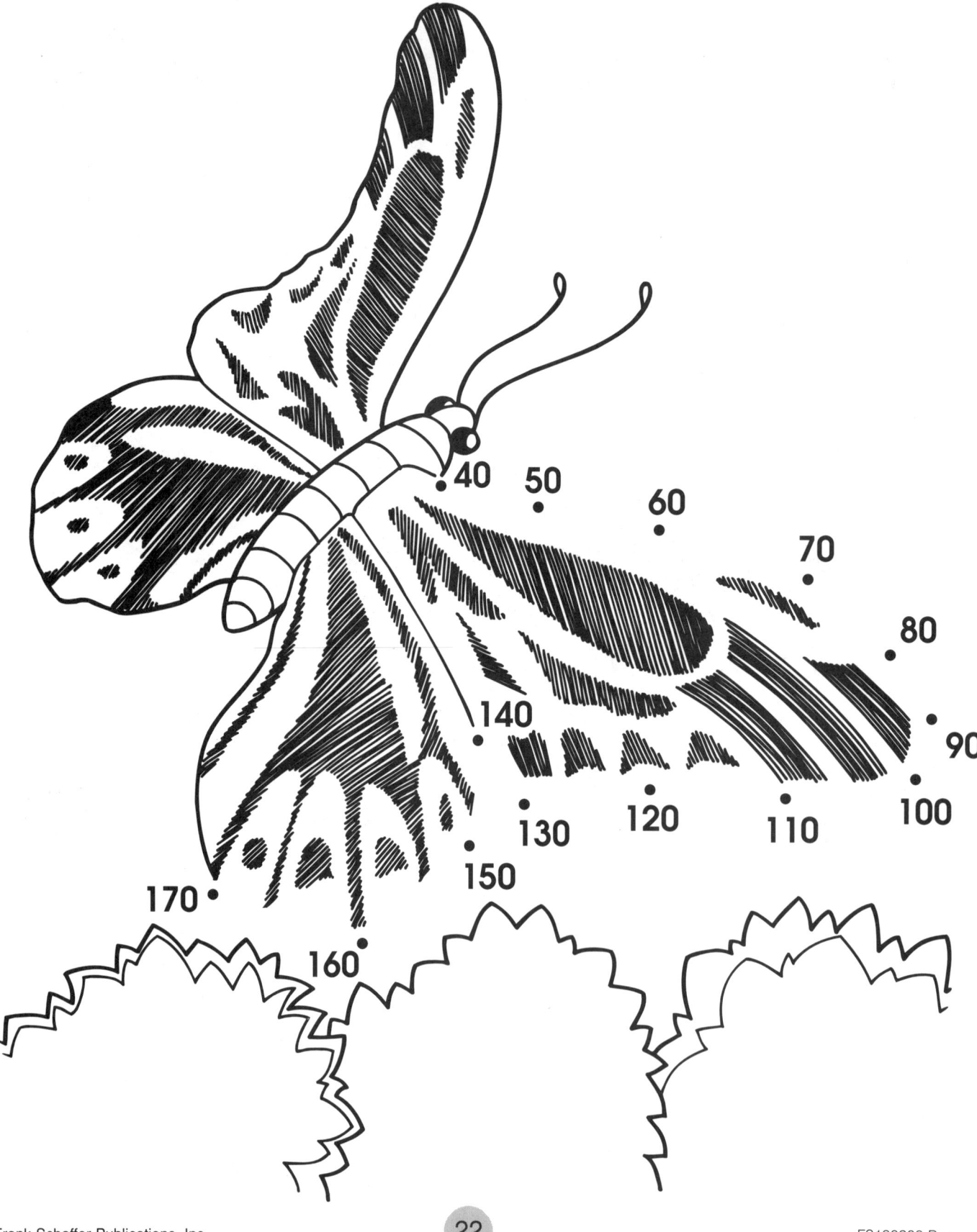

Greg the Grasshopper

Name _____ Skip counting 30–180

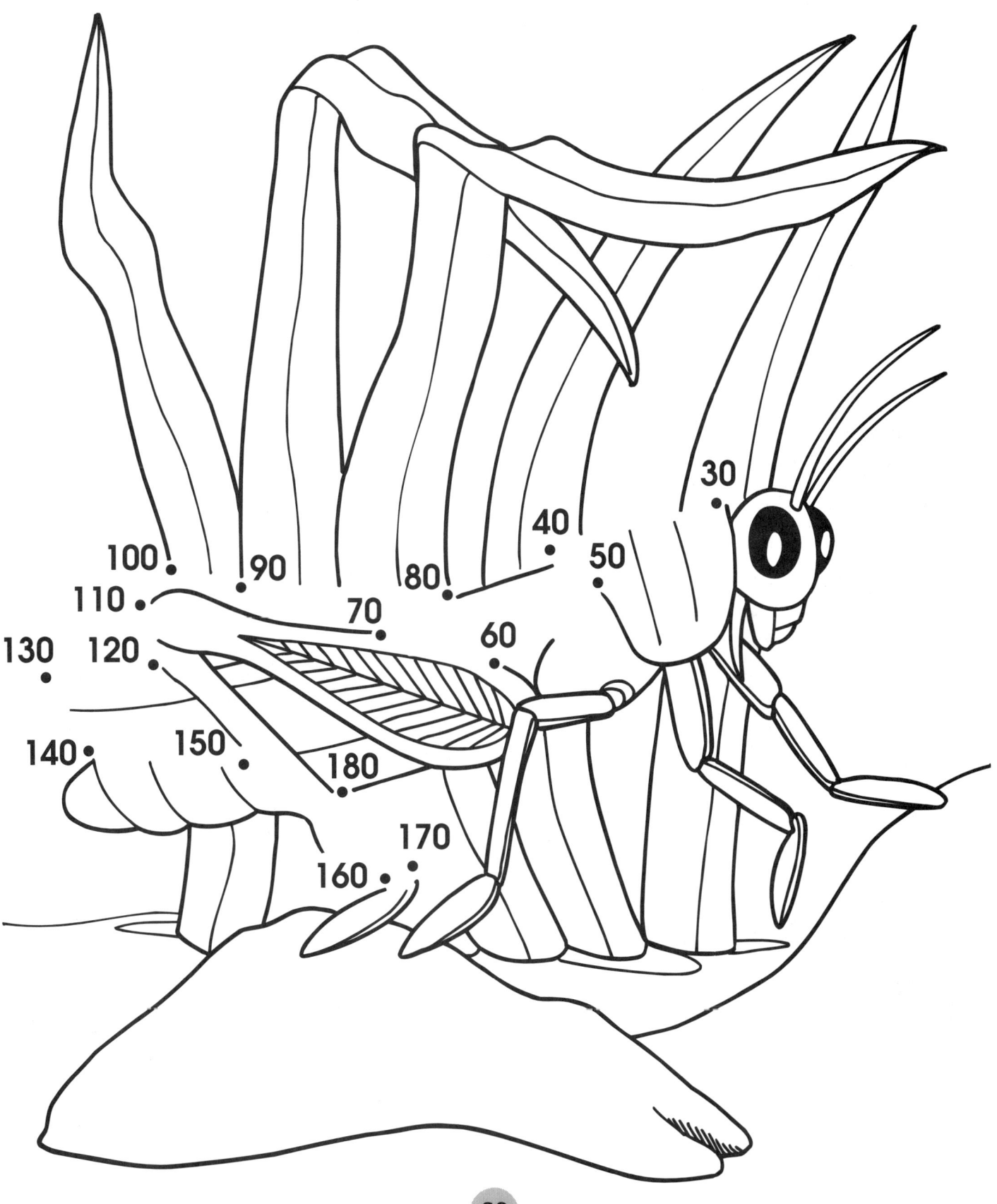

Scotty the Stag Beetle

Name _____ Skip counting 50–200

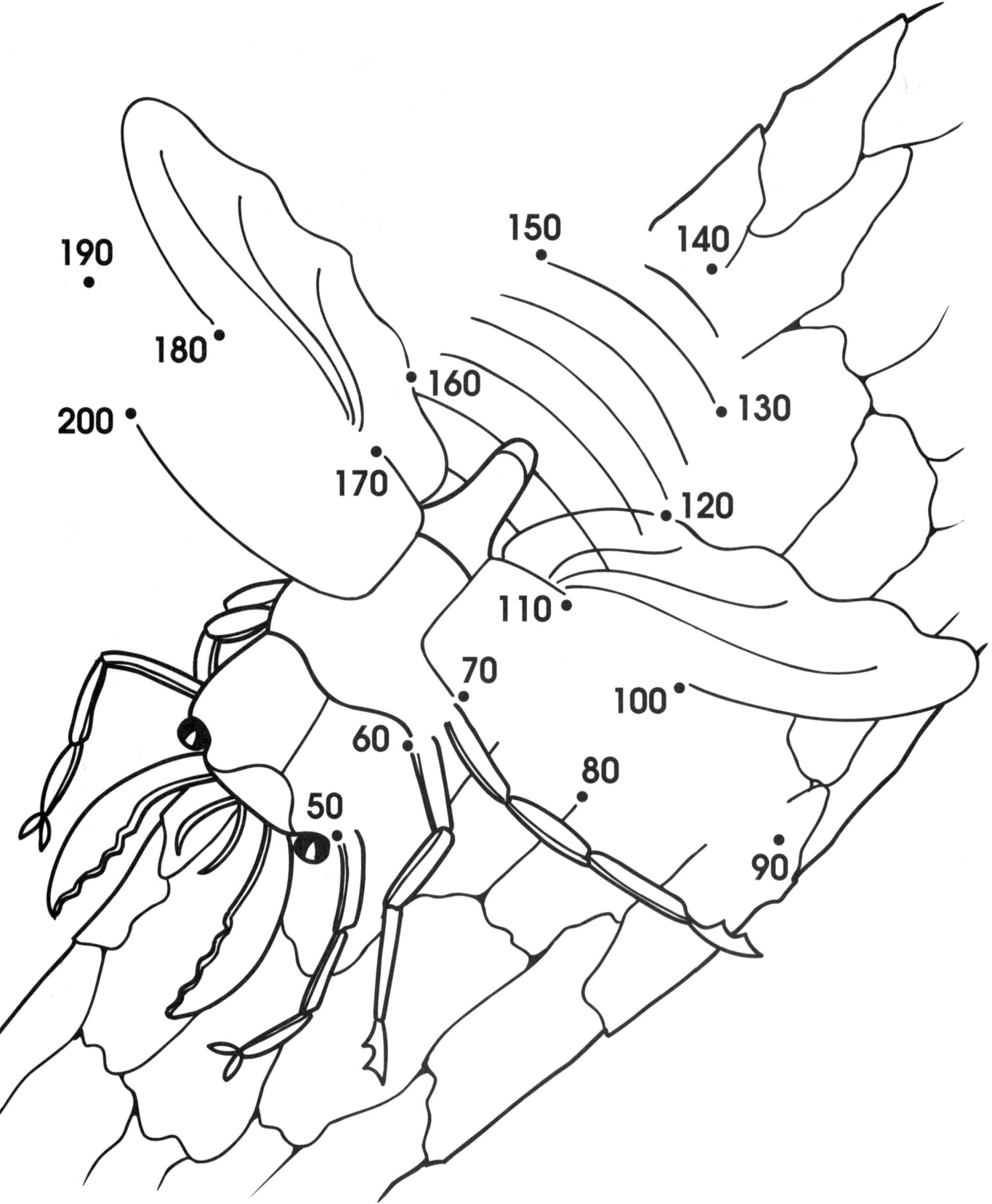

Sara the Swallowtail Butterfly

Name _____ Skip counting 0–85

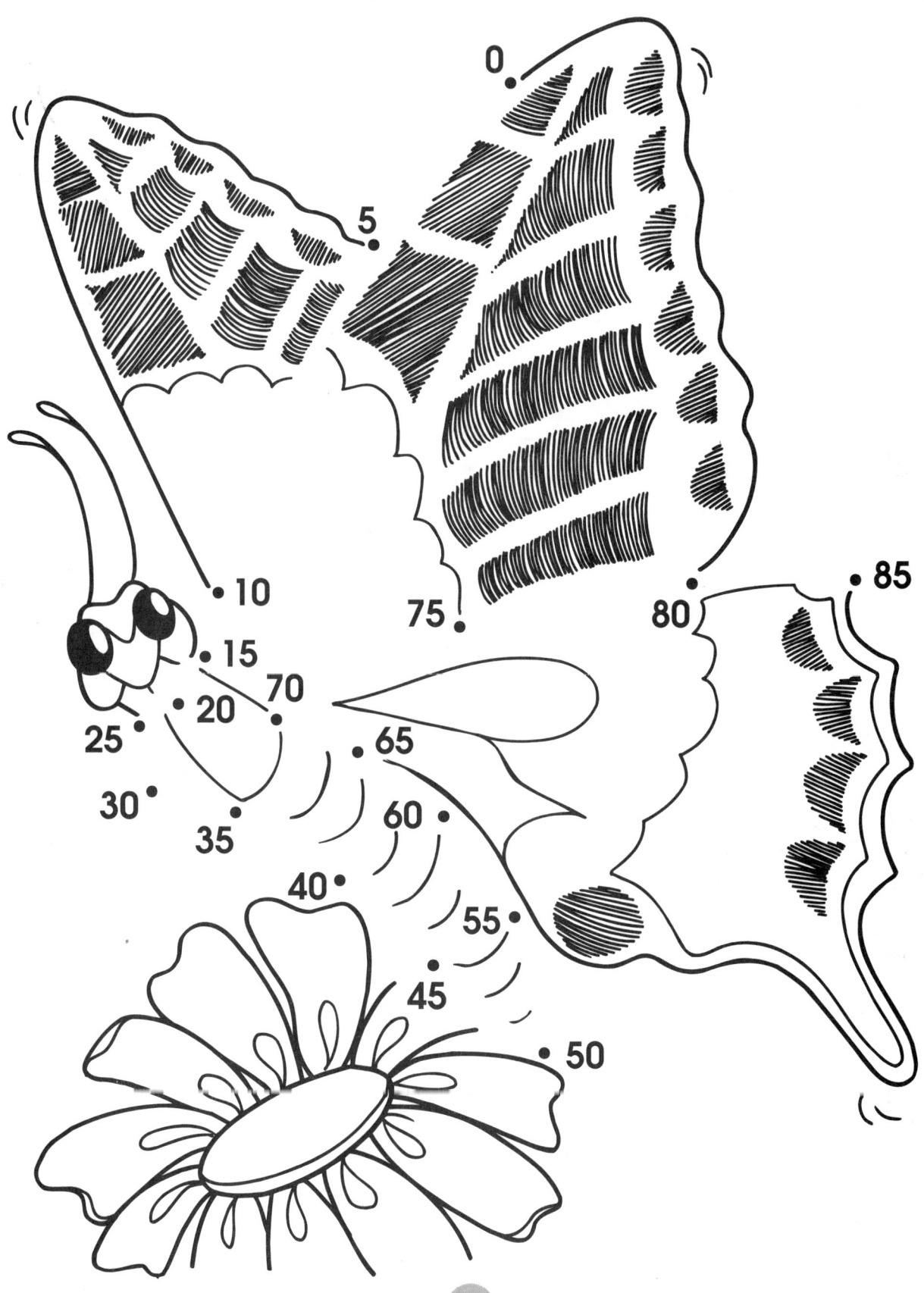

Chris the Cricket

Name _____

Skip counting 15–100

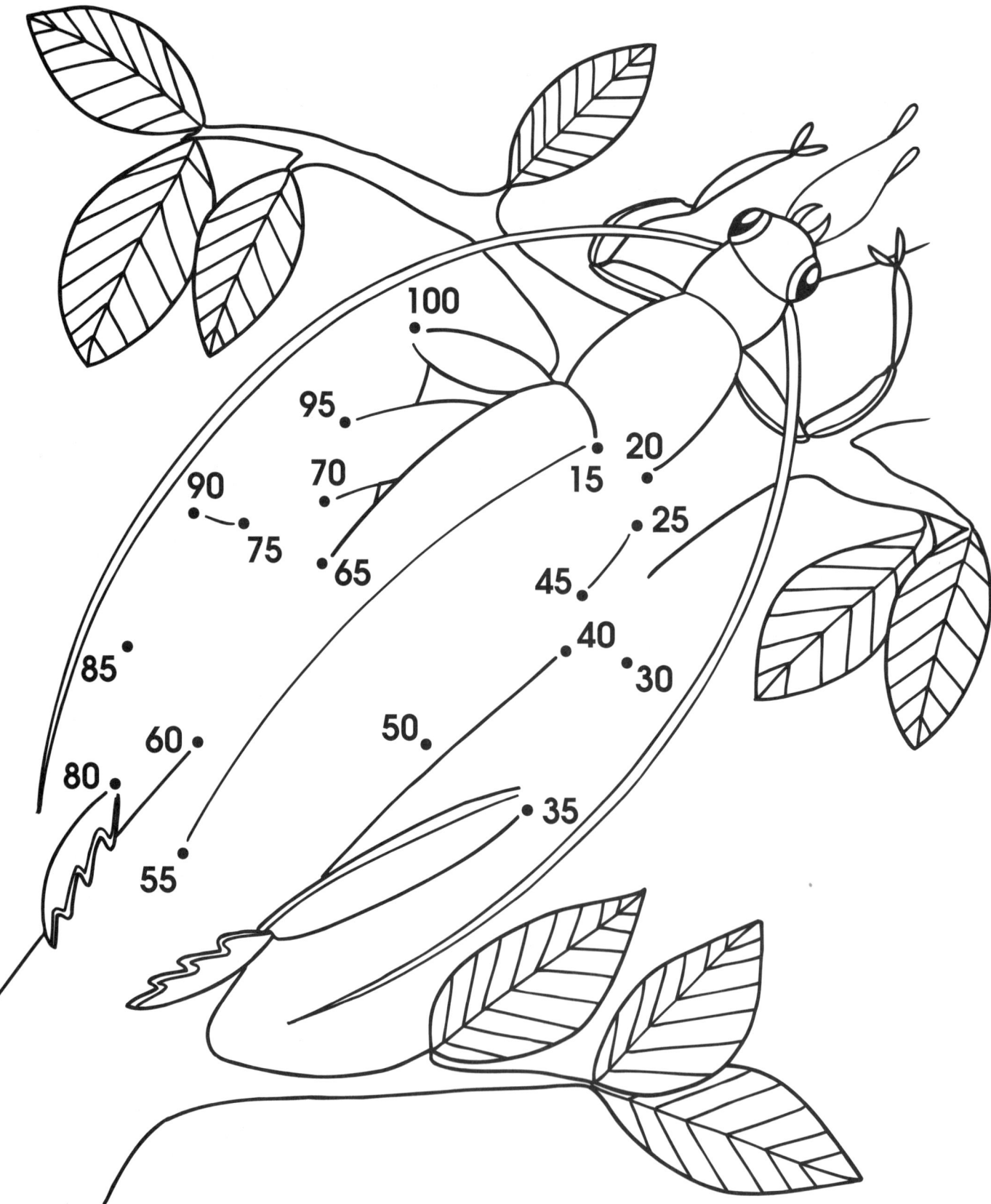

Lori the Leaf Insect

Name _____ Skip counting 0–95

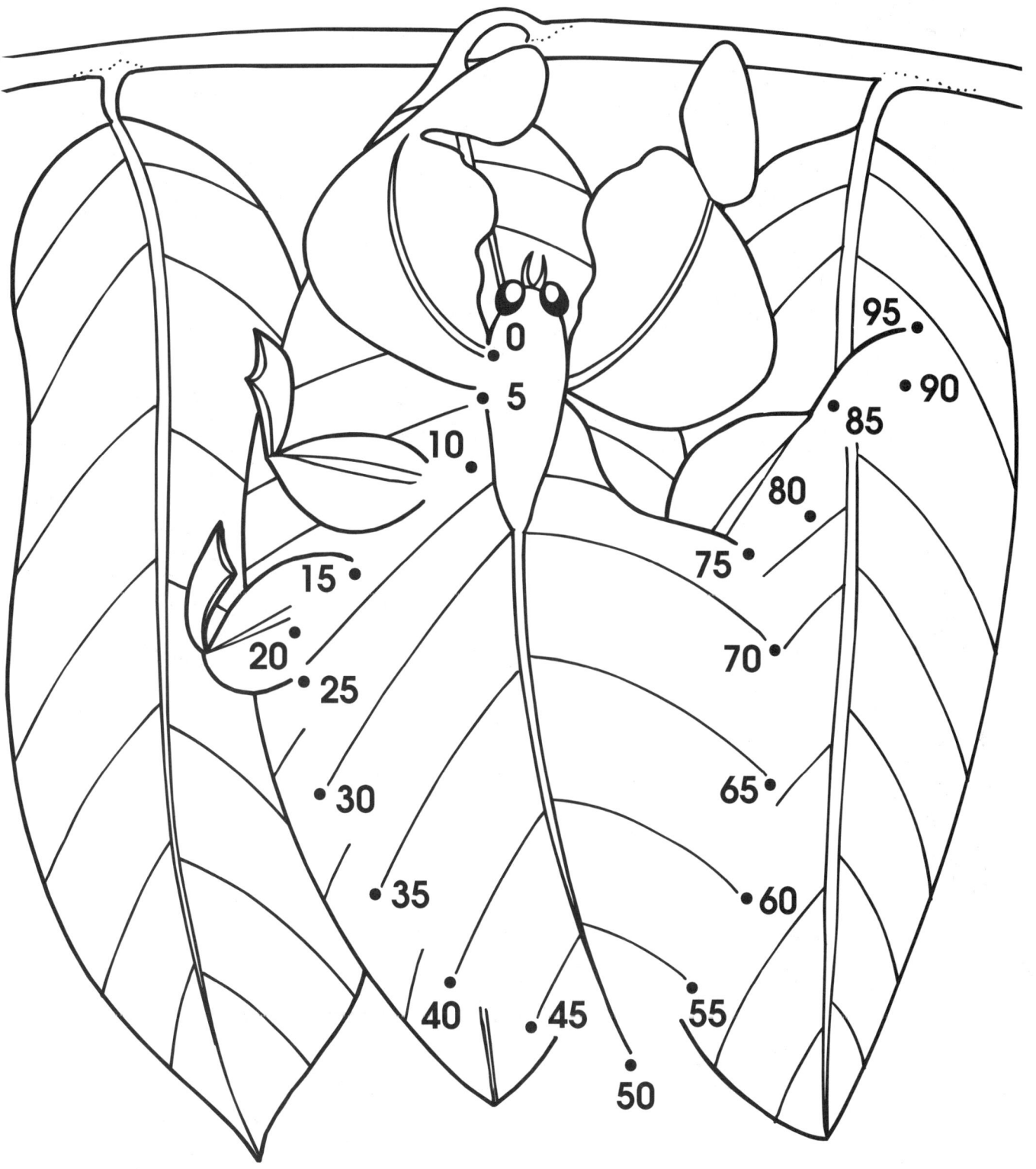

Helen the Housefly

Name_____

Skip counting 0–100

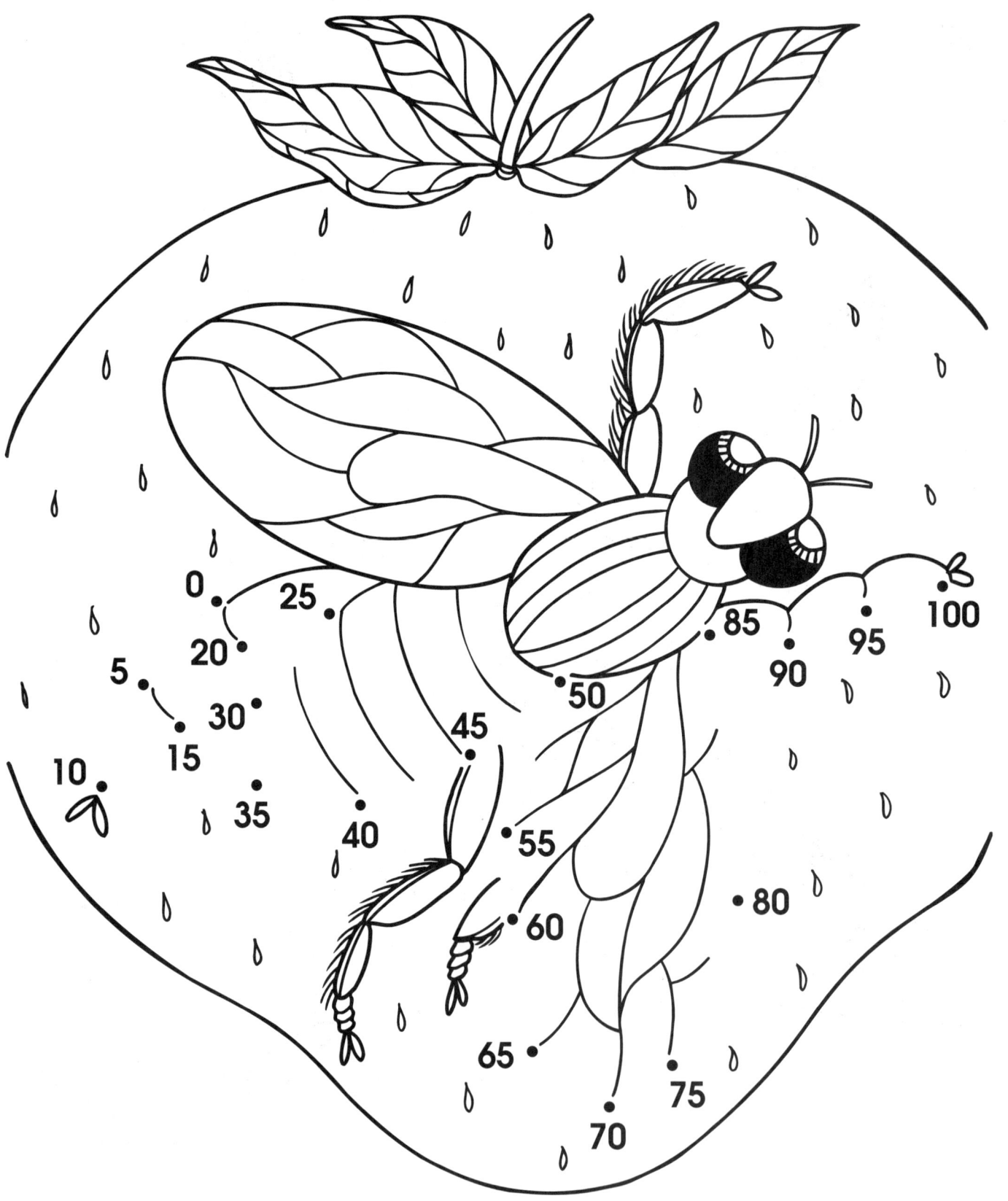

28 reproducible

Hal the Hercules Beetle

Name _____   Skip counting 0–170

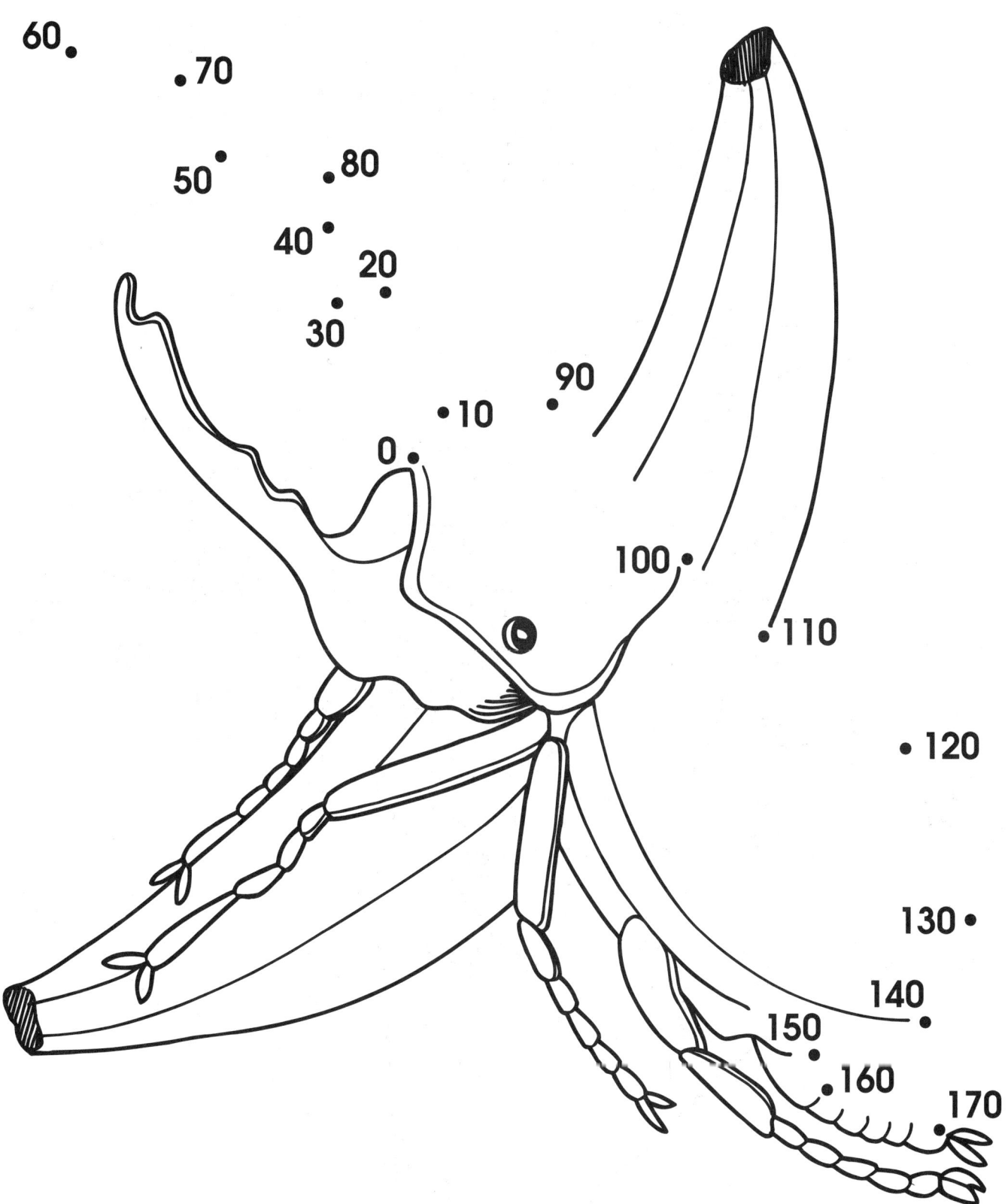

Carrie the Caterpillar

Name _____

Skip counting 20-190

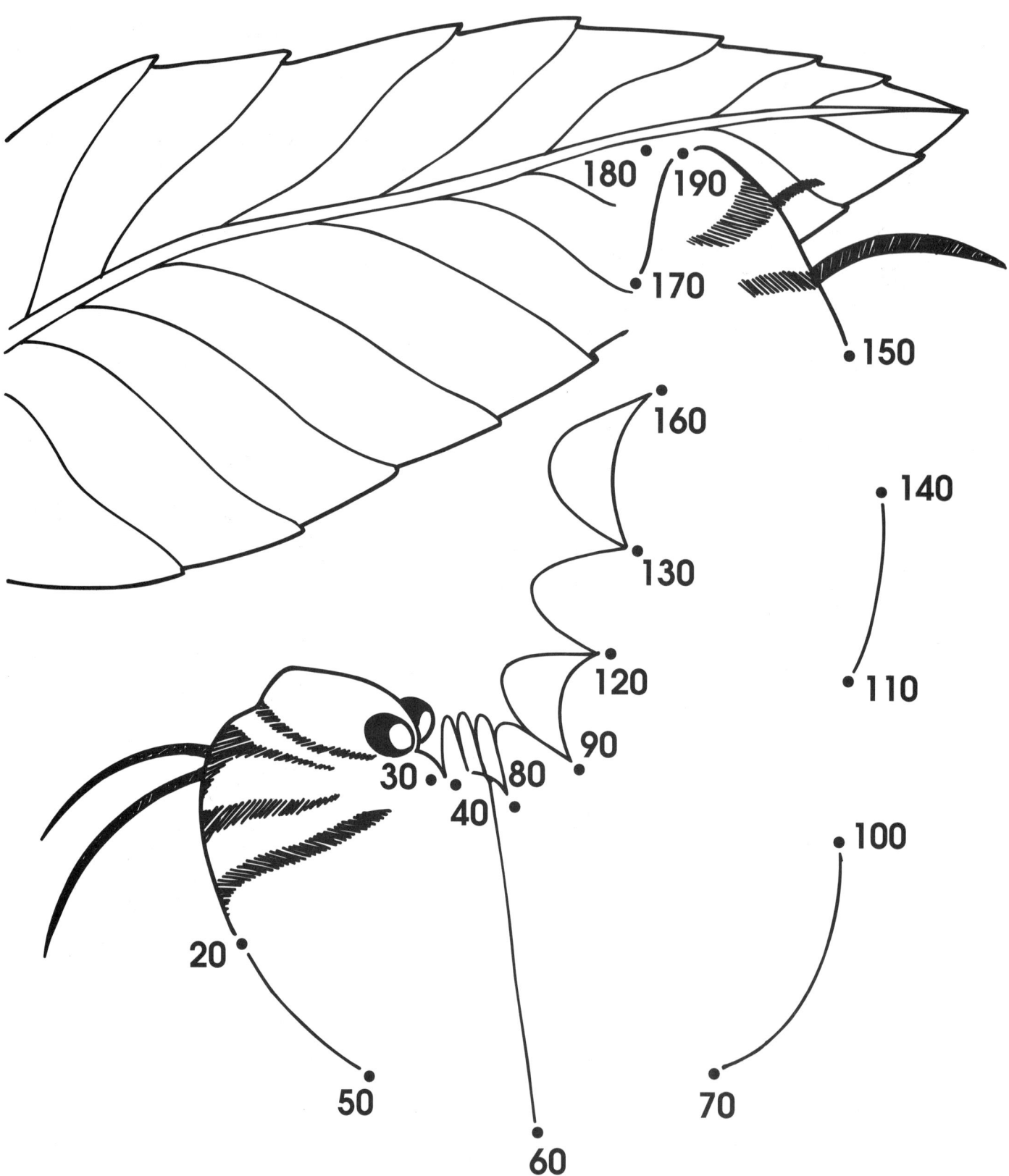

Lynn the Luna Moth

Name _____

Skip counting 0-190

Barry the Beetle

Name _____

Skip counting 10–200

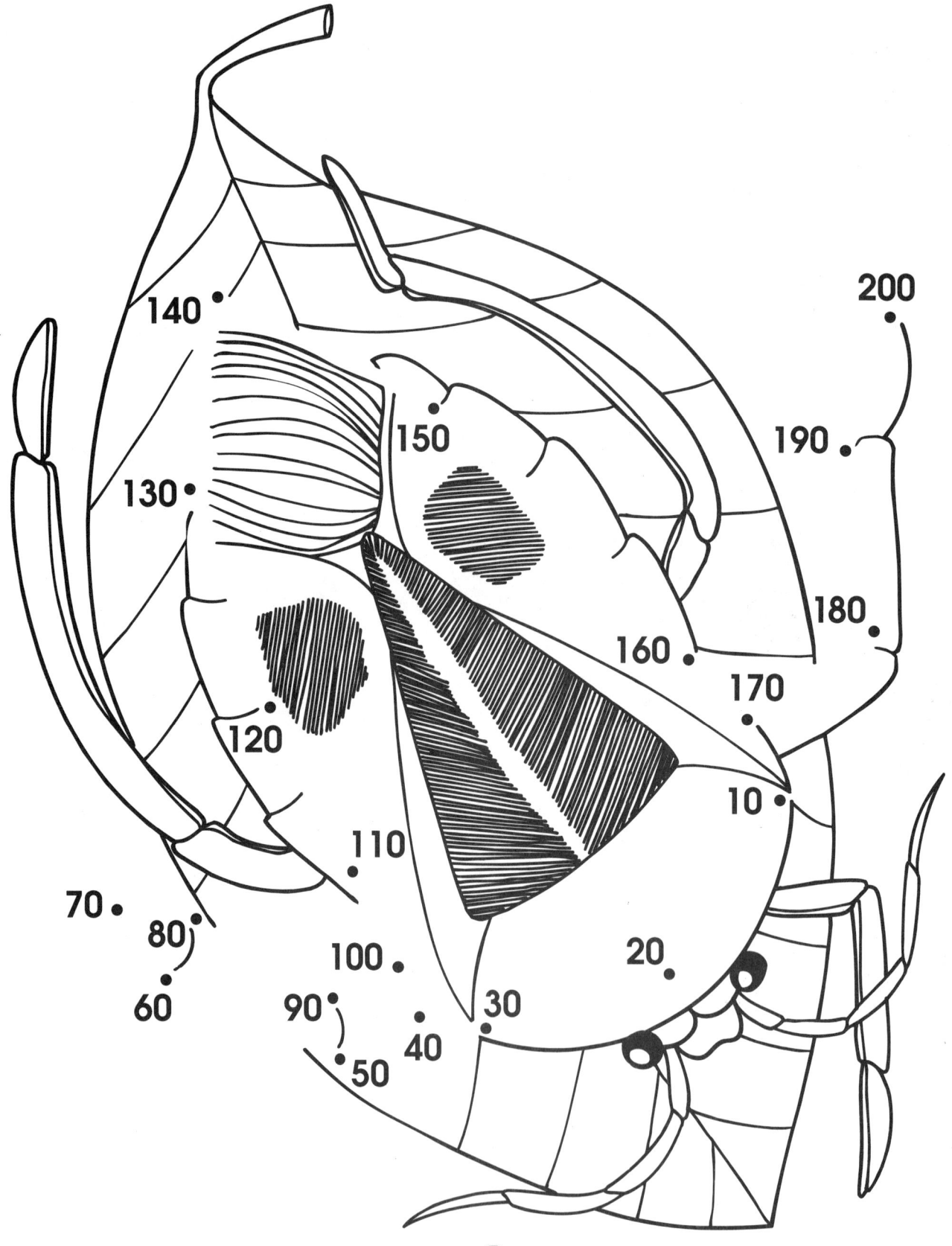

© Frank Schaffer Publications, Inc.

FS130308 Bugs